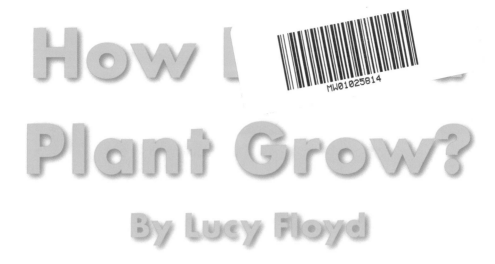

How I Plant Grow?

By Lucy Floyd

Copyright © by Harcourt, Inc.

All rights reserved. No part of this publication may be reproduced or transmitted in any form or by any means, electronic or mechanical, including photocopy, recording, or any information storage and retrieval system, without permission in writing from the publisher.

Requests for permission to make copies of any part of the work should be addressed to School Permissions and Copyrights, Harcourt, Inc., 6277 Sea Harbor Drive, Orlando, FL 32887-6777. Fax: 407-345-2418.

HARCOURT and the Harcourt Logo are registered trademarks of Harcourt, Inc., registered in the United States of America and/or other jurisdictions.

Printed in Mexico

ISBN 978-0-15-363615-8
ISBN 0-15-363615-7

2 3 4 5 6 7 8 9 10   805   16 15 14 13 12 11 10 09 08

SCHOOL PUBLISHERS

Visit *The Learning Site!*
www.harcourtschool.com

 a small seed

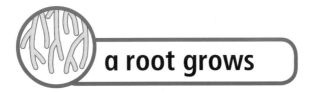 a root grows

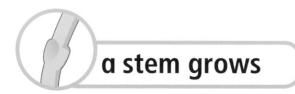 a stem grows

 leaves grow

 new plants